# HOW TO MAKE A SILENCER FOR A .22

PALADIN PRESS
BOULDER, COLORADO

*How to Make a Silencer for a .22*



ISBN 0-87364-759-9
Printed in the United States of America

Published by Paladin Press, a division of
Paladin Enterprises, Inc., P.O. Box 1307,
Boulder, Colorado 80306, USA.
(303) 443-7250

Direct inquiries and/or orders to the above address.

# Contents

## Warning

Since 1934 silencers have been legal in the United States only if they are registered to their owner and a $200 tax has been paid on their transfer. Current federal statutes (the Omnibus Crime Control Act of 1968 and subsequent regulations) allow a maximum penalty of 10 years in prison and a $10,000 fine for the illegal manufacture or possession of an untaxed, unregistered silencer for a portable firearm. The agency of enforcement of these federal laws is the Bureau of Alcohol, Tobacco, and Firearms of the U.S. Treasury Department. Enquiries regarding the procedures for applying for a license to construct a silencer and for paying the appropriate tax should be directed to that office.

In addition, many state and local jurisdictions have restrictions on the possession of silencers, even if appropriately taxed and registered with federal authorities.

The information presented in here is *for informational purposes only.*

Part One

# How Silencers Work

Before we attempt to make or use any silencer, it behooves us to learn how it works and why it works that way. This knowledge can be invaluable if, at some later date, we must come up with a field-expedient version or we must make a tactical decision about suitability of a particular "can" (silencer) for some proposed task.

## WHY A .22?

First let's clear up some of the common misconceptions about silencers. A silenced .22 is not the ideal weapon that novelists and gunshop commandos would have you believe. It is really suitable for squirrel-sized game only. Precise placement of a .22 bullet into the brain is required to drop a 150-pound animal. As you can imagine, this is much easier said than done.

A rifle can deliver such performance, but I have seen world-champion pistol shooters miss a 5-yard head shot when under stress. The .22's little 40-grain slug carries only about 100 foot-pounds of muzzle energy when it is held to subsonic velocities (as it must be if we seek the maximum in sound reduction). In order to be certain of the outcome of a body shot on a heavy animal—and not have to rely on luck—at least 300 foot-pounds of energy must be transmitted to the target. Such performance is beyond the capability of any pistol cartridge smaller than the 9mm Luger (or Parabellum, as it is known in Europe). If the bullet overpenetrates the target, whatever energy it carries with it out the other side is wasted, at least in relation to the immediate goal.

Multiple hits do help, but not nearly as much as you might think. A second hit might increase the probability of a stop by 10 percent. A 9mm hit might have a 60-percent chance of achieving the goal, but another such hit would improve the odds only to 70 percent.

A .22 slug is made of soft lead and has very little momentum to keep it driving on its intended track. Both of these factors limit the .22's penetration. A .22 has been known to fail to penetrate even a rib, and the skull at the frontal plate of most game animals is made of much heavier bone. The angle at which the bullet strikes the target affects its penetration, and if this angle deviates from the ideal 90 degrees, you should not expect a .22 to penetrate. The brain is a small target measuring about 4 inches across, and the head is a highly mobile part of an animal.

Having said that, I still advise you to give some thought to other factors. A novice should never begin with the most difficult problems in any endeavor because of the likelihood of making mistakes while learning.

Thus, the beginner should start with a $30 used single-shot rifle, not an expensive .45.

It is the ease of manufacturing and mounting a silencer on a .22 that accounts for the fact that most silencers are found on these little rimfires, not any special lethality of the cartridge. The .22 is the most commonly encountered unsilenced firearm, as well, because of its low recoil, excellent accuracy, and low cost both for the gun itself and its ammunition. The .22 typically has an exposed barrel, which simplifies silencer attachment a great deal. Cans have been mounted in many ways, including by glue, tape, clamps, solder, welds, threads, and press-fits. Of all of these methods, only threading deserves serious consideration. Improper soldering or welding on a barrel can ruin it, and the other methods are generally either too insecure or too imprecise in their alignment of the can with the barrel.

## THE PHYSICS INVOLVED

For a can to truly work, it must be precisely aligned with the bore (the inside of the barrel). If this alignment is disturbed (or was never present) because of poor construction or abuse, it will cause a disastrous loss of accuracy and can even be dangerous to the shooter! Using a silenced firearm as a club or dropping it on a hard surface is almost certain to ruin this all-important alignment, so suppressed arms must be handled with care. This is one reason that the world's military does not make more use of suppressed armament.

Silencers can improve accuracy by reducing the shooter's tendency to flinch (i.e., jerk at the trigger, a subconscious reaction in anticipation of the muzzle blast). As can be imagined, anything that will increase hits should be seriously considered by the individual combatant (whether group planners deem it worthy or not). A good can will also eliminate muzzle flash, which can be of obvious value in nighttime engagements.

Cans work by delaying the exit of the hot powder

gases into the air. In order to do this, a can must be capable of momentarily containing these gases or reducing their volume by cooling. The high-pressure gases will take any path open to them, so a silencing effect can be achieved by constructing many dead-end pathways inside the can, while providing a clear tunnel for the bullets to follow through and out of the can.

These powder gases are what propel the bullet out of the barrel and they also work the action of most autoloading arms, so we cannot eliminate them from the equation. The muffler on your car works in much the same way as does a silencer. However, the muffler is ordinarily mounted 6 to 8 feet away from the heat source. This distance allows the gases to cool considerably in the exhaust pipe. A silencer is mounted just scant inches away from the heat, so the steel wool and fiberglass employed in muffler construction do not make good silencer baffling. These materials will burn out in two or three shots, and this is not acceptable performance.

While you may hope that one shot will accomplish your objective, you would be extremely foolish to rely on this. A can that will not stand up to at least two magazines full of rounds being fired through it is worthless. Why do I say this? Because it is necessary to fire one magazine full through it to test for reliability before embarking on the actual undertaking, and you should have one magazine of service left with which to handle any unforeseen complications. Cans alter where the gun shoots, i.e., where the bullets impact on the target relative to where the sights are aimed. Knowing this, we realize that every time the can is repacked, it must be retested. Where does this leave you, if you must repack every three rounds? Up the creek, that's where.

The correlation between the sights and the shot's point of impact must also be checked each time the can is remounted on the gun or if it is subjected to any trauma. This inconvenience makes it desirable to just leave the can on the gun. This is known as "dedicating the piece to silenced use." Given the bulk of most cans and the need for discretion implied in silencer use, you can see that silencers can be a mixed blessing in the field.

I have attempted to make all of my designs as compact as the latest technology will allow, as well as easy for the average guy to make and mount. Most of them will last for hundreds of rounds, and all are easily cleaned or repacked.

Once we know what makes a silencer work, we can draw on that knowledge to infer other things about it. For instance, if Cartridge A contains more powder than Cartridge B, we can be fairly sure that A will require a larger can to contain the gases adequately. However, this is not to say that a direct proportional relationship exists between case volume and can size, and especially not between straight-walled pistol cases and the bottle-necked rifle cases. A .223 may burn six times as much powder as a .45, but that does not mean that the .223 needs a can with six times as much internal volume. This difference is attributable to several factors. A rifle barrel is three or four times as long as a pistol barrel, and this allows the gases to cool a bit. But rifle powders also burn more slowly than pistol powders, so this is a confused issue. The fact that the neck of a rifle case is constricted appears to slow down the gases' exit, and this may account for the apparent discrepancy in the size of the cans.

If a piece has a high cyclic rate, it will require a larger can to handle the increased volume of gases generated. This is particularly true of the 20-shot-per-second MAC-10 and the Micro Uzi because the can is still pressurized from one round's gas when another is fired. In addition, high-powered rifle rounds and submachine guns generate more heat, and this puts an additional load on cans intended for such guns. These cans must have a heat-disperser built into them at their rear, so that the regular baffles are not subjected to this additional heat. These two requirements mean that cans for these guns must be significantly larger than semiauto pistol cans.

You may have heard the old wives' tale about using a baby-bottle nipple as a silencer or the one about the Irish potato. By now you should have an accurate idea as to how effective such devices are.

Another myth is that of the "silenced revolver." The slightest leak of gas anywhere in the gun-and-can assembly is enough to make a lot of noise. All revolvers have a huge gas leak between the front of the cylinder and the rear of the barrel. Thus merely screwing a can on a revolver barrel will not silence it (nor will holding a pillow in front of it). Either technique will work on semiauto pistols, but one can hardly sight through a pillow, and the use of sights is necessary for consistent results!

A silencer must be a pressure vessel. The bullet serves as a partial plug at the front of the can at the moment of firing. The tighter the fit of the bullet to the hole in the can, the more effective the can will be because it will do a better job of containing the powder gases. Geometry tells us that if we want a close fit between the slug and the hole in the can, we must use great care to see that the can is mounted concentric to the bore. This means that not only must the can be on the barrel straight, but also that the parts of the can must be made and put together with a minimum of tolerance or "play" between them. If the bullet strikes anything inside the can, bad consequences can be expected.

A bullet must be spin-stabilized in order for it to fly straight. The bullet has this spin imparted to it by spiral grooves that are cut into the interior walls of the barrel known as the rifling. The marks that they impress into the bullet are what the ballistics lab can use to determine if a slug came out of a particular gun.

A bullet is made to fit fairly tightly in the gun barrel and must upset (expand) into the firing grooves to make a gas-tight seal for the gun to work efficiently. If the powder gases could slip past the bullet, a great deal of velocity would be lost and the barrel would be subject to accelerated erosion. The bullet must upset upon firing, so that the necessary seal is formed. When the bullet upsets thusly, it then takes the rifling grooves properly and is set to spinning on its long axis, which stabilizes it for its flight to its target. The rifling grooves are said to be "engraved" into the bullet as it is forced through the bore by the expanding powder gases. This is one reason that a bullet must be made of a material that is softer than that of the barrel.

The bullet will only fly true and straight for as long as it is undisturbed. If the bullet were to strike one or more of the baffles inside the can, it would be disturbed to some degree. A 230-grain copper-jacketed .45 slug can tolerate more of such a happenstance than can a soft-lead .22 slug weighing only 40 grains.

The rate at which a bullet spins varies. This rate of spin is determined by the rate of rifling twist, which is measured in turns per inch. A .45 might be said to have a twist of 1-in-16, meaning that the rifling would make one complete turn in 16 inches of barrel. Naturally, if the bullet is driven to a higher velocity, it passes through the bore more quickly and thus will also be driven to a higher spin rate.

If a can is not properly aligned with the bore or if

the baffles are not properly made, the bullet will strike whatever is in its path. Depending upon the severity of the blow, the bullet may suffer only a mild upset, or it may become completely destabilized. If the latter occurs, it will flop end over end upon its emergence from the can. Such a flopping bullet is said to be "keyholing," the name being derived from the characteristic elongated hole such a bullet makes in a paper target. A keyholing bullet is extremely inaccurate. It cannot be relied upon to hit a gallon jug at 20 feet! Also, a keyholing slug's penetration capabilities will be seriously reduced. Since the .22 is not exactly a great penetrator under the best of circumstances, a keyholing .22 lacks much tactical value.

A further problem that can arise is that the bullet may begin keyholing while still in the can. As you may imagine, the slug will then encounter even more resistance and can even become lodged in the can. To say that this is embarrassing is to understate the case considerably. You can now understand why just stuffing a bunch of steel wool into a tube doesn't cut it in the real world. If the target can be approached within a few feet, why bother with a gun, especially an underpowered .22?

When Hollywood shows us a typical cinematic silencer, one for a 9mm is about 2 inches long, and its outside diameter (OD) is about an inch. Would that it were so! The smallest 9mm can commercially available (to the military and police only) measures 5 1/2 x 1 5/8 inches. Attach this to a typical 8-inch 9mm pistol and you've got a fair-sized portability problem. If the common cylindrical configuration is used, my best 9mm design measures 5 x 1 1/4 inches. It is possible to reduce the length to 4 inches, but it requires sophisticated construction techniques that I will not go into here. It is also possible to shorten a good 9mm pistol to 6 1/4 inches, resulting in a package that is only 10 1/2 inches in overall length (OAL).

At any rate, you can readily understand why most people in the field settle for a .22. A typical .22 silencer will run to 6 inches in length and be 1 inch OD. Add this to the typical 7-inch .22 pistol, and the result is just barely shoulder-holsterable.

There is another reason why powerful pistols are not commonly silenced by tinkerers, and, to understand it, an explanation of auto pistol construction is necessary. The low-powered cartridges (.22, .25, .32, and .380) have correspondingly low chamber pressures, on the order of 15,000 pounds per square inch (psi). A pistol for these calibers can have an unlocked breech (i.e., the slide is held shut on the cartridge only by the inertia of the slide and by the power of the recoil spring). A 9mm, on the other hand, develops 30,000 psi and must have some sort of locking or delaying mechanism incorporated into its design if it is to be of a handy size. Submachine guns do not need this delay mechanism because they have large, heavy breech bolts (whose inertia effectively holds the breech closed) and strong springs.

The form that this delay mechanism takes is usually the Colt-Browning lock (used on Colt-Browning, SIG, Star, S&W, etc.). In this design, the barrel both tilts and moves back and forth a short distance. Locking lugs on top of the barrel mate with corresponding grooves machined into the inside of the slide (when the slide is forward against the chambered round). Before the slide can release the barrel from its locked position, these lugs must disengage. This disengagement is effected by a camming surface on the underside of the barrel which bears when the slide moves rearward under recoil. If more than a 9-ounce weight is attached to the end of the barrel, the slide will not recoil sufficiently and will not unlock—and you have a manually operated firearm! If a powerful auto pistol did not to have some sort of delaying mechanism, it would have its slide blown open prematurely when the chamber pressures were still too high. The result might be a face full of hot gases and brass fragments for the shooter, and the piece would quickly batter itself to bits.

I have a design for these autos that does permit normal cycling of the action and does not interfere with the sights. However, it requires all-aluminum construction and, hence, heli-arc welding. By employing this design on a modified .22, I can achieve a package that is only 6 1/2 inches OAL. This is small enough to be carried in, and fired from, a wallet holster.

My design for the .45 auto results in a 12-inch OAL and can be worn in a shoulder holster that is both fast and discreet.

The above-mentioned examples are just some of what can be achieved with my designs, but I must recommend that you gain some familiarity with a simpler design before attempting these more complicated types.

## HOW MUCH SILENCE CAN YOU EXPECT?

The best material for baffle construction is window screen wire, but it can't be just any screen wire. The aluminum and fiberglass types can't withstand the heat encountered in silencers and will melt if you try to use them. I have used brass, bronze, copper, and stainless steel screen to good effect, but bronze is easiest to work with and is usually more readily obtainable than the others, so I recommend it. Small-town hardware stores do not normally carry it, however, because it is five times more costly than aluminum screen. It does require annealing before it will form proper baffles, but that is a very simple matter to take care of.

Screen wire folds and spindles very neatly and easily, which is a big help in baffle construction. I know of cases where makers have tried to use strands of copper wire or copper scrubbing pads and the results leave a lot to be desired, plus it is a lot more work than doing it my way. Once the screen is compressed into the doughnut configuration called for in this manual, the thousands of tiny interstices serve as perfect traps for the hot gases to expand into. A further advantage of screen wire as baffle material is that all the tiny wires absorb a lot of heat from the gas, which of course reduces its volume and helps avoid muzzle blast and its associated noise.

Some World War II silencer technology called for the stacking of hundreds of single-thickness screen-wire washers inside a tube, and this sort of baffling is still being recommended (by others). While this method does work, there are a couple of problems with it when Joe Average tries to employ it. Unless you just happen to have a punch press and the appropriate dies, you are stuck with cutting out the washers with a pair of scissors and then forming the central hole with a leather punch. Even if you take the shortcut of making square washers, this is still a very tedious process. By contrast, my method allows you to form and assemble the six to eight doughnuts to the can in only 20 minutes with a simple hand-powered die set that anyone can make in an hour or two, using only hand tools.

The potentially more serious drawback of the stack of washers method, though, is that the loose ends of wire around the central hole can become liabilities if they come unraveled, as they are wont to do under the heat and blast effects of repeated firings. If enough of those wires protrude into the path of the bullet, the previously mentioned ill effects can be expected. World War II designs as developed by Bell Laboratories under contract to government agencies used a special heavy-gauge wire screen with a tight weave and strong crimp of the warp and weave, which did not unravel. Similar screen is now used in the pulp industy to carry pulp through the paper and hardboard presses.

I employ two sizes of baffle in this design. One is longer than the other, and it also has a larger central hole. These two sizes of baffle are arranged alternately in the can so that the gases will expand into the larger hole in the large baffle and then be forced to contract in order to pass through the smaller hole in the small baffle. This method exposes more of the baffling material to the gases, resulting in more efficiency and a smaller can. Other designs sometimes employ a similar strategy, but they use only a piece of tubing as a spacer between their baffles. This is less efficient because the spacer has no absorptive capability.

If the maximum in sound reduction is desired, a firearm with a sealed breech is necessary. By this I mean a bolt-action, lever-action, pump-action, or break-open single-shot. I say this because a small amount of gas exits the ejection port of an autoloader along with the fired casing, Some silenced autos incorporate a slide-lock (a manually operated lever that serves to hold the breech of an auto closed until the firer works the slide by hand). In view of the fact that the noise that comes out of auto pistols is only a mild puff of sound, I do not recommend giving up the fast repeat shots that you may need! A semiauto can hit at the rate of five shots per second, but it would take at least a second to work the slide manually.

Some makers of (overweight) cans for powerful

pistols actually cite their failure to cycle as an advantage because they make a little less noise. They also think that you don't need to use the sights since their design blocks the sights. This alone is enough reason to doubt their expertise because 30 years of competition has taught us that sights must be used for consistent hits when under stress. A 6-foot head shot is easily missed by the so-called "point-shooting" method and so is a 15-foot chest shot.

On a piece that fires full-auto, this ejection port noise can be objectionable, especially on the high-cyclic-rate types exemplified by the MAC-10. You may have read about the noise made by the bolt's movement and that which comes from the primer's firing, but these two noises are completely submerged in the ejection port noise.

There is a way to effectively muffle the ejection port noise of most submachine guns. It involves the use of a product called a "brass-catcher." This is a hopperlike contrivance that can be clamped over the ejection port of most popular guns and serves to keep the ejected casings from being scattered all over the countryside and lost. An outfit named Rigid markets one made of plastic that serves our purpose here very well. If the joint between the mouth of this hopper and where it contacts the receiver is sealed gas-tight, no noise can escape. Use silicone caulking compound to achieve this seal. Of course, once the hopper is full, it will have to be removed, or it will cause failure-to-eject malfunctions.

Since we know that the blast of the escaping gases is what causes the noise, it follows that the smaller the hole in the can, the more effective it will be at its job of suppressing noise.

One source of noise that is often overlooked is

that which is made when the bullet strikes something. Depending on what is struck, this sound may be only a soft thud, but if metal is hit it will, of course, ring like a gong. Naturally, the more powerful the cartridge, the louder this noise will be. Other factors that will have some bearing on whether this impact noise is noticed are how many such noises are made and how close together they occur in time. If a car is shot up with a silenced subgun, for instance, it sounds as if someone was using an air hammer on it.

A variation on the above concern is how much noise a failing or flopping target makes. A great deal depends on the background setting involved. A rural area at midnight is typically much quieter than a busy city street. So if the wrong ears are likely to be in range, a quieter piece would be called for.

One of the least understood aspects of silenced arms involves the use of supersonic ammo. Many people are not aware that, in addition to the muzzle blast that results from the high pressure gas following the bullet out of the bore, the bullet makes a sound if it exceeds the sound barrier. No silencer can do anything at all about this so-called "sonic crack" since it is caused by the shock wave that precedes the bullet as it flies to the target. This shock wave impacts the ground, trees, buildings, etc., and sounds like someone popping a bullwhip. This cracking noise is normally obscured by the muzzle blast, and it is audible only to about 200 yards in open country when a good suppresser removes all of the sound except the sonic crack. A normal, unsilenced shot, by contrast, can be heard for miles, especially over water or in canyon country.

Conversely, some people think that the bullet going supersonic is what makes guns noisy. If this were true, a .45 auto would be silent because typical ammo for the .45 achieves only about 800 feet per second (fps) velocity. The speed of sound is about 1,090 fps, so the .45 is a subsonic round.

If we desire the utmost in silence, we must restrict ourselves to the use of subsonic cartridges. A fact that is not commonly known is that the velocity of a round can vary from shot to shot by as much as 70 fps, even if the rounds are taken from the same box of ammo. This means that we cannot crowd our load much over 1,000 fps or we risk having an occasional round "crack" on us.

If we are restricted to subsonic ammo, we cannot reach out much past the 100-yard range, at least not very practically, because the bullet is falling so quickly.

Targets in the field do not often allow spotter shots, although silencers can help in this regard. When using subsonic .22 ammo silenced, I have had animals just duck when the slug whips past them, seeming to look all around themselves as though trying to determine what sort of hornet is after them! I have taken up to five birds off a line, right beside each other, within just a few seconds. Survivalists should appreciate the ramifications of this facet of silenced firearm use. In time of crisis, when every round is treasured and every opportunity to obtain food is to be seized, a silencer can be every bit as valuable as the guns and ammo.

Since almost all calibers exceed the speed of sound when fired from a rifle, we must ordinarily resign ourselves to the fact that we will have sonic crack when we employ a long gun. Even regular .22 Hi-Speed long rifle (LR) rounds achieve 1,200 fps when fired in a rifle. The trick here is to use LR target ammo, which does remain subsonic in rifle barrels. Your local discount store is not likely to carry these specialty cartridges; you will probably have to go to a real gunstore to find them. This ammo costs more than the regular stuff because it is made to higher standards of quality control. Rimfire target shooters discovered that rounds that stay subsonic tend to be considerably more accurate in the .22, so they leave the high-speed stuff to the plinkers.

Oftentimes poor silencer performance is passed off as sonic crack. In order to properly evaluate a can's performance, you must set up a test that eliminates as many extraneous factors as possible. To do this, fire into the air to keep bullet impact noise out of the picture. Get away from any buildings or trees that the sound wave can crack against. Hold the bolt shut to eliminate ejection port noise. On low-powered rounds, simply holding a gloved band against the rear of the bolt will suffice; on the bigger stuff, a padded C-clamp or vise grip will do the job.

The worst myth about silencers is that they are used only by murderers. This is pure nonsense put out by people who can't think of any other use for guns either. There are millions of guns out there, and only a few thousand are ever used for criminal purposes. The fact that their guns were loud did not stop any of our notorious assassins, and clubs and knives are used to commit more crimes than are committed with guns. A criminal will simply find another way to achieve his ends. It may be explosives, poison, or silk stockings, but—since he gets to choose the time and place—no amount of legislation can stop him. The Swiss have a full-auto assault rifle in every household, and permits for silencers, rockets, even mortars are not hard to obtain. Yet they have a very low crime rate. What makes the difference is who has these devices, not the equipment itself.

The design put forth in this manual is the best on the market both in terms of efficiency of operation and ease of construction and mounting on the gun. It is the result of a great deal of research and anguish. You may rest assured that it works if you follow the instructions carefully. Read through the entire book at least once, making certain that you understand all of the principles involved before taking any drastic steps.

Part Two

# Construction Techniques

I sought a way to make a quality silencer without access to either a lathe or a welder. Although you will see references to the use of these tools, this can is quite easy to build without them. The dimensions are different when using PVC and epoxy because a 3/8 x 24-inch bolthead will not fit into a 7/8-inch OD steel tube, which I have found to be the best size if one does have access to a lathe and welder.

The male threads are put on the can instead of on the gun as in most other designs. This was done to eliminate hassles. If you don't have a lathe, how would you thread the outside of a barrel and get it straight? How would you create the shoulder that must be present for the rear of the can to abut so that it is square with the bore? How would you deal with remounting front sights, a barrel that can't be removed from the receiver, or a barrel that is covered by the slide? By the simple step of putting the male thread on the can, you can whip all of these problems.

The counterboring of the gun muzzle requires that you have a piloted drill bit even if you do have access to a lathe. A regular drill bit will not follow the bore exactly, so do not blame this design if you try to get by without making this special drill bit. The reduced-diameter nipple on the nose of this bit can be ground by hand on a bench grinder. The brass pilot is necessary even if you can precision-grind the nipple on a tool-post grinder. The hardened steel of the drill bit will ruin a barrel if it is spun in the rifling ahead of where the counterbore is to be cut.

Another nice feature is the use of a washer and an internal spring as the end cap for the can. Since a can must contain pressures, the front of the can must be sealed off in some manner. If it were not, all the baffles would be blown out at the first shot. Most designs call for threading the inside of the silencer tube, making a matching threaded end plug and screwing it solidly into place. This requires a lathe and considerable skill and wastes a lot of time. It is also dangerous, and so is any other type of solidly mounted cap. Remember what I said about silencer alignment? If a high-powered round should strike a solid cap, what do you think would happen? Since the internal groove in this design is only 0.025 inch deep, such a mishap will just cause the washer and snap ring to be ripped out of the can, rather than causing a blowup right in front of your face. This groove can be cut neatly with only a Dremel hand-held electric grinder or an electric drill when using PVC tubing.

## MAKING THE BAFFLES

I cannot give exact sizes for the dimensions of the rectangles of screen wire that (after folding, spindling, and forming) will become the baffles for this design because different materials, mesh, and wire size will compress to different degrees. Also, people's ability or willingness to fight the screen into the forming die will vary. This all assumes that the builder will take the trouble to obtain the exact OD and wall-thickness tubing recommended here, which will probably not be the case nor need it be. Once you understand the principles outlined in the next paragraph, you will be able to adjust the size of the rectangles to your specific needs.

The length of the rectangle of screen will be governed by the amount of the stuff that you can spindle around the rod and still get the assembly started into the cylinder of the forming die. Since this design calls for two different hole sizes, two different sizes of rod must be used to form them. You will not be able to get as much screen wrapped around the larger rod and still stay within the size required to get into the cylinder, so the large-hole baffles will require a shorter rectangle of screen to begin with. The width of the screen rectangle is what determines the length of the finished baffle (or its thickness, if you prefer to think of it that way). Therefore, the larger baffles will require a rectangle that is shorter and wider than the rectangles for the small baffle.

If the above seems a bit confusing, just go over it a few times and refer to the illustrations, Figures 6-16. The screen is first folded lengthwise and then spindled around the rod. Simply cutting out a rectangle of paper, folding it, spindling it around a pencil, and starting the package into a toilet-paper roll should suffice to show you how it is done.

Bronze screen wire has too much elasticity to form a good baffle as it is, so it must be annealed. If this simple step is overlooked, the baffles will not have their holes properly aligned and they will tend to unravel in use. Annealing will also make the screen easier to spindle and form.

The annealing process is very simple and takes only 10 seconds per rectangle, so there is no reason to skip it. Just fold the rectangle into thirds lengthwise, hold one corner in a pair of pliers, and pass the screen slowly through the flame of a kitchen stove or a propane torch. Do not stop moving the screen while it is in the flame or you will burn a hole in it. Once the entire rectangle has been made to glow red-hot, allow it to cool, and the job is done. The screen cools very quickly and can be handled with bare hands in less than a minute.

I stated earlier that aluminum screen could not be used for silencer construction. However, it can be used to make the silencer's front baffles (only) if the other materials are in short supply. The aluminum screen does not need to be annealed before use.

### *Rubber Disk Wipes*

This design employs rubber disks that are made out of inner-tube rubber. They are simply cut out with the same scissors used to cut the screen wire, after tracing a circular pattern on the rubber (see Figures 24 and 25). No attempt should be made to cut the central holes in these disks (or wipes, as they are called). Any attempt to do so will actually hurt accuracy. It is impossible to tell where the center of the disk will be after it is assembled into the can because it must be slightly oversized to obtain the necessary seal and will be compressed to one side or the other as it is pushed home into the tube. After assembly, getting the hole in the right place is simply a matter of firing a shot and then burning out the hole cleanly by passing a red-hot rod through the baffles and the rubber disks.

The use of wipes in a design is somewhat controversial, since they do contact the bullet as it travels through the can, and the potential for disturbing the slug's stability does exist. However, the demands put upon a can vary, and what some squirrel hunter regards as marginal accuracy for his rifle is considered superlative accuracy in a training pistol. Some seek small size in a can; to others bulk is irrelevant. I was able to obtain 2-inch groups at 25 yards using a bull-barreled Ruger .22 target pistol that employed wipes. That design was completely concealed inside the gun's normal barrel configuration, and to achieve the necessary efficiency wipes were necessary. If the wipes' material and thickness are correct for a particular type of can, why not just go with what works?

The technique of burning the central hole after assembly helps accuracy by ensuring that the bullet is not struck more heavily on one side than another, and this may account for my success. For rifle use, I recommend that the wipes be burned out to the full size of the hole in the baffle (rather than leaving them undersized as in a pistol can) in the interests of maximum accuracy and because bulk is not likely to be an issue on a rifle. It is easy to adjust the size of the hole in the wipe by simply pausing a bit longer at each one that you want to be larger when passing the heated rod through the wipes. Even if the wipes do not touch the bullet, they can still add to a can's efficiency by compartmentalizing the can, thereby stopping the linear flow of gases through the screen wire baffles.

The maker must decide early on whether he will use PVC pipe or the metal tube because that decision will have a bearing on some of the other dimensions involved in the construction of the can.

The person interested in the ultimate in sound reduction will want to have a 0.235-inch hole in the baffles, but such a small central hole requires great care in all aspects of the work. It will require that the adapter be welded into the tube first, and then be finish-cut, threaded, and drilled while still chucked in the lathe. A mandrel should be used to keep the silencer tube from being distorted by the lathe chuck. The part of the baffle-forming die set that I call the "piston" (Figures 13 and 14) is just the right size to serve as this supportive mandrel. The piston is also the right size to use as a template for making the rubber wipes.

Some of the tiny pocket autos cannot tolerate having the barrels counterbored because they are already on the borderline of not having enough rifling to adequately stabilize the bullet. If 3/4 inch of rifling is lost to the counterbore, their barrel is ruined. A particularly bad example of this is the Iver Johnson TP-22. The later models of the Jennings are okay (since they went to Microgroove rifling), but they will require the use of 5/16 x 40 threads on the can because their barrels are only 3/8-inch OD. The little Beretta M21 is marginal but is such a sweet design that I developed a method especially for it using a replacement barrel liner to extend the barrel to 3 inches and to provide an exposed surface on which to cut an outside thread.

## THE TUBE

You will notice that the dimension given for the silencer tube varies from 4 to 6 inches. This is because the PVC tubing is 1/8-inch thick. Since its OD is 3/4 inch, that leaves an ID of only 1/2 inch. A silencer must have a certain amount of internal volume in order to do its job. If the inside diameter (ID) is reduced, the length must be increased. The reverse doesn't hold true, unfortunately. Increasing the OD doesn't let you shorten the can, at least not very much. The 4 x 7/8-inch can is for short-barreled pistols and will need cleaning after only 30 rounds.

A rifle will need the 6 x 7/8-inch can. Substituting the PVC tubing for metal will necessitate adding 2 inches to either length. The fact that a rifle needs a longer tube for the same cartridge may seem strange, but it is caused by the larger plug of air present in the longer barrel. This air takes up can volume at the instant of firing. If the gases cannot expand into the baffles (because they are full of compressed air), they will follow the bullet out of the can, resulting in noise.

My baffles are easily cleaned, so you need not always completely repack the can when carbon buildup starts to clog up the baffles and make them inefficient. Simply soaking the can in carburetor-cleaning solvent and blowing it out with a compressed air nozzle will restore it to service. In a pinch, gasoline can be substituted for the solvent.

In view of the tyrannical nature of all governments, everyone should at least know about the efficient use of firearms and their accessories. In view of the current restrictions on silencers, isn't it nice that screen wire can be found lying around everywhere? The same holds true for bolts, drill bits, inner tubes, and tubing. The survivalist who cannot be content with having only the materials and the tools handy should bury his "toys" somewhere, not on his own property. The site should be shielded from prying eyes, and the materials should be stored below the frostline. Freezing and thawing can crack the pipe. CPVC is recommended over PVC for storage containers because it was developed specifically for burial and comes in a wide variety of sizes along with matching caps and a special glue to seat them. Including a few packets of silica-gel desiccant would be a good idea to keep the goodies from rotting or rusting.

## DRILLING

Making the necessary "piloted" tap drill is a fairly demanding task. If you have never sharpened a drill bit, you will need to acquire some experience at that before attempting the more complex job of regrinding the cutting lips of this bit after creating the 3/4-inch long .170-inch OD nipple on its front end (Figure 30).

As a drill is used, the edge of the cutting lips becomes worn and chipped. As this condition worsens with use, the increased friction caused by the blunted edge will speed up the deterioration of the cutting edge until the bit eventually refuses to cut at all or until it seizes up and breaks in the hole.

When the machinist sharpens a drill bit, he grinds back the area just to the rear of the cutting edge to restore the proper cutting qualities to the bit. The actual edge ordinarily would receive only a light touchup grind unless the bit has been seriously abused. In the case of recreating the cutting lips on this piloted bit, quite a lot of metal must be removed.

Look carefully at a new drill bit. Notice that the chisel point comes exactly at the center of the bit (i.e., the length of the two cutting lips is the same). The lips diverge from the point at the same angle: 59° (Figure 21). Industrial supply stores stock a drill bit angle gauge that is an aid to achieving this angle. This gauge can also be used to check the length of the two lips on an ordinary drill bit.

With a piloted drill bit we cannot use the simple angle gauge to check the length of the lips because the nipple obscures where the actual center of the bit is. We can use the gauge only to determine the angle of the lips and then make test cuts into a piece of scrap metal to check the cutting action. These cuts should be made at very low speed so that the action of the lips can be observed carefully. If one lip is removing more metal than the other, the one doing the most work must be ground back to some extent. How much this lip should be ground back is something that must be determined by a process of cut and try. With experience, you will become a better judge of how much metal to grind off.

Another thing to watch for is whether the entire length of the cutting lip is working. When the bit first starts into a hole, only the part of the lip closest to the center will be cutting. After the full diameter of the hole is started, however, the full length of the cutting lip should be engaged in removing metal.

Ideally, these test cuts should be made on a piece of old .22 barrel. Call around to your local gunsmiths and check to see if they have a short section or even an entire barrel that got rusted out or bulged somehow. Do not expect to get it free; you'll have to pay about $10.

If a piece of barrel is not available, just use a #3 drill bit to drill a 0.213-inch hole in a piece of waste stock. A .22's bore diameter runs about 0.210 inch (compared to its groove diameter, which is 0.224 inch). The groove diameter is measured from the bottom of the rifling and thus does not affect the clearance of our pilot as we drill the counterbore that will permit us to cut the female 3/8 x 24 threads into the muzzle of the gun. In order for the drill to turn, the brass pilot must be about 0.005 inch smaller than the bore, or about 0.205 inch.

Of course, we cannot make our test cut without the brass piece in place, so let's look into that. Obtain a 3/4-inch-long piece of 1/2-inch threaded brass rod, or the same length of 3/8-inch unthreaded brass rod. Drill a hole (the same size to which you reduced the nipple of the drill tap—approximately 0.170 inch) through this brass rod lengthwise. If a lathe is not available, you must indicate the center of this piece so that you can drill it with a band drill. To do this, color the end of the piece of rod with a Magic-Marker. Measure the OD of the piece with a vernier caliper. Set the vernier at one-half of that measurement and hook the movable jaw of the venier over the edge of the piece of brass. By drawing the (fixed) jaw of the vernier over the inked face of the brass, the center of the piece can be marked. Center-punch the indicated spot and use a 1/8-inch center-drill to start the lengthwise hole. Repeat this process on the other end of the brass piece. Then drill all the way through the piece with a 1/8-inch regular bit and follow this up with whatever size bit matches the nipple you have ground on the tap drill (Figures 28 and 29).

If the nipple was ground by hand rather than with a tool-post grinder in a lathe, the nipple can be expected to vary a bit in diameter and to be a bit off center as well. All this may mean that you need to drill a slightly larger hole in the brass than what you think. As long as there is no hardened steel exposed when the brass is reduced to fit the .22 bore, all will be well.

The best way to mount the brass piece on the nipple is to silver-solder it on, after degreasing and fluxing both parts. However, if a torch is not available, Acraglas-Gel epoxy will hold it on. This amazing stuff is available from Brownells, the gunsmithing supply people in Montezuma, Iowa. Acraglas chemically heats itself after it is mixed and it sets up like a rock after about 20 minutes, so you must read the instructions and get everything ready before you start.

After mounting the brass piece, its OD must be reduced to a slip-fit in the bore. This can be done by spinning the bit in a drill while holding the brass against a bench grinder. Once the OD of 0.205 inch is achieved, the test cuts can be done. (Finish-grind the nipple in this way, too.)

As I stated earlier, the main thing to look for in the test cut is the openness of the two cutting lips' working action. Slowly depressing the drill's trigger will allow you to observe the chip removal and thus to gauge whether the bit is ground correctly. Refer back to a new bit, if necessary, to remind yourself what the correct relief angles look like, until you've got it right.

## MAKING THE BAFFLE-FORMING DIE SET

Let's take up making the baffle-forming die set as a training exercise. It may seem like a put-down, but it's better to screw up on this than on counterboring the barrel or installing the threaded adapter in the rear of the can!

The die set consists of five parts:

1. Cylinder (Figure 1)
2. Piston (Figure 2)
3. Die-washer (Figure 5)
4. Bolt (or small rod) (Figure 3)
5. Large rod (Figure 4)

The **cylinder** is nothing more than a 3-inch piece of the same tubing that the can is to be made of. The cylinder's ID must be the same as that of the can, or the baffles formed in it will not fit in the can. Do not be tempted to use a cylinder of slightly smaller ID to make it easier to insert the baffles into the can because the baffles' central holes will not be concentric to the bore. Do not try to form the baffles in the can itself. If you do, the baffles at the rear of the can will be too tightly compacted, while the front baffles will be too loose.

The **piston** is a piece of solid stock 3 inches long, reduced in diameter so as to be slip-fit inside the cylinder. This piston has a 1/4-inch hole drilled lengthwise through its center. This hole is then counterbored to a larger diameter most of the way through the body of the piston. This counterbore allows the use of this piston with either of the two rods, so that we can form the two different-sized baffles.

The **bolt** or small rod is made out of either a 3-inch piece of 0.235 inch (15/64) rod, or it is formed by cutting the threaded part off a 1/4-inch bolt that is 4 inches long and has only 1 inch of threads on it. Your hardware store will have such a bolt, but you may have to go to an industrial supply place to get the 15/64 rod (a good reason to stick with the 1/4-inch hole in the baffles). If you decide on the smaller hole, get enough rod to form the 9-inch-long wipe-burning rod while you are at the store.

The **large rod** is a 3-inch long piece of solid stock that has a reduced-diameter nipple formed on one end. This nipple enables us to use the same die-washer with either the bolt or the large rod. If forming this nipple is a problem because you do not have access to a lathe, simply make another die-washer and have one for use with each rod. The large rod is not used for the PVC version.

A word about making **die-washers** the right size (Figure 5). A lathe is naturally a big help, but I promised that this could all be done without one, and so it will! The ID of a washer can be drilled larger while it is held in a vise, so start with fender washers, which come with undersized holes.

## REDUCING THE OUTSIDE DIAMETER

Reducing the OD is tricky, and I recommend that you take it to someone who does have a lathe. After all, he won't have any idea what it's for. Make it easy for him by mounting the washers on a bolt that is a close fit to the hole in the washers and which is threaded all the way up to the bolthead. Secure the washers with a lock washer and a nut. He can then center-drill the bolthead and mount it in the lathe on a live center in the tailstock. Tell him to use a sharp lathe bit, high RPM, and light cuts. Otherwise the washers will spin on the bolt, and achieving an accurate cut will be difficult.

If you must reduce the OD of a washer without a lathe, you need a vernier caliper, a bench grinder, and some sort of coloring agent. DyeKem layout blue works best for this, but you can get by with a Magic Marker. Color one face of the washer. Calculate how much metal must be removed from the OD. Divide this figure in half. Set the vernier at this figure and use it as a scribe (as outlined in the paragraph on drilling the brass pilot, pages 27–28). When you are done, you will have a line marked all around the circumference of the washer. Lock the washer onto a bolt as described above and chuck the bolt into a drill. Hold the drill up to the bench grinder, with the bolt pointed downward at the floor. Holding it thus will reduce its tendency to spin on the bolt, as well as allowing you to see the scribed line.

The decision about whether to use PVC tubing must be made at this point.

If a lathe is not available, another way must be found to reduce the OD of the piston (Figure 2) so that it is a slip-fit inside the cylinder. The actual dimension is not critical as long as a loose wire cannot get caught between the piston and the cylinder as a baffle is being formed.

One way to reduce the piston's OD is to draw-file it. To do this, clamp the piston horizontally in a vise, with some of the piston's surface exposed above the top of the vise. Stand at the end of the piston. Hold a

12-inch flat bastard file flat and horizontal in both hands, and draw it back toward you while bearing down on the top of the piston. This will cause a ribbon of material to be peeled off the top of the piston. Move around the circumference of the piston, repeating this step and reclamping the piston as necessary. This is a quick way of removing quite a lot of material by hand.

If your drill will not open up enough to accept the 7/16-inch rod, making the nipple on its end will have to be done by hand. This is no big deal because the large size of the hole formed with this rod will permit quite a bit of error before the hole is enough out of line to have any effect, so you can just use the bench grinder for this step.

Once the parts of the baffle-making die set have been made, they should be tested by making some baffles, which should then be measured and tested for correct size and form. Figures 6-16 show various stages of this process.

## FORMING BAFFLES FOR FIT

We begin with a rectangle of bronze screen wire. This screen is folded in thirds lengthwise (Figure 6) and then annealed, before starting to form it into a baffle. Actually, thirds isn't quite accurate since the center section of the screen must be wider than the other two sections so that when the screen is folded the loose ends of the wires are not at the edges of the folds.

Once the rectangle is folded and annealed, it is ready to be spindled around whichever rod the maker selects. Some auto pistols (notably the PPK) will not function reliably with a can unless the rear baffles have a large hole, making a sort of expansion chamber just in front of the muzzle. With that in mind, let's run through the steps in making a large-hole baffle.

The screen will compress differently when

annealed than when left as is, so when trying to determine the correct size of the screen rectangle, do not make the mistake of setting the size with unannealed screen.

Note that the screen is spindled around the large rod with the loose edges to the outside (Figure 7). This is to eliminate the possibility of their coming unraveled in the path of the bullet. A further consideration is the loose wires present on the end of the folded rectangle. To remove these loose ends, fold approximately 1 inch of the end of the rectangle back upon itself, rough edge to rough edge (Figures 8 and 9), and begin the spindling from the resulting fold rather than from the cut end (Figure 14). Note that the rectangle for the large baffle will be about 3 1/2 inches wide by 4 1/2 inches long (if using the 7/8 inch OD tubing). The actual size must be determined by trial and error. Start with an oversized rectangle so that the excess can be trimmed off until the spindled screen (and rod) can be started into the cylinder. By "started" I mean that the screen is of a size to permit its entering the cylinder if its edges are worked into the cylinder with the ends of the pointed paper scissors that I recommend for cutting the screen. Working it in is the technique of starting one part of the wire roll's circumference into the cylinder and then moving around the cylinder, tucking the screen into the end of the cylinder the way you would put a tire on a wheel by hand.

If you try to shortcut this step, you will find that your baffle hole is not in the center of the cylinder. While this might be acceptable for the large baffle with its larger central hole, it would definitely lead to trouble with the smaller baffle and its closer tolerance. I recommend that all of the baffles be made by tucking the screen into the cylinder in the interest of adequate compression to ensure that the baffle does not become unraveled. This is especially important if you are not going to solder the ends of the baffle after forming it.

Once the screen-and-rod assembly is started into the cylinder, just push it in the rest of the way to where the nipple on the large rod almost emerges from the other end of the cylinder. Slip the die-washer into the end of the cylinder and around (onto) the nipple. Stand this assembly on the bench, washer end down, and make certain that the washer is fully inside the cylinder. Now slide the piston into the upper end of the cylinder, with the large hole of the piston slipping around the end of the large rod (Figure 14). Strike the protruding end of the piston a solid blow with a rubber mallet and the screen will be formed into a large baffle. Remove this baffle from the cylinder and slide it off the rod.

These baffles hold their shape pretty well, and you can now check the length. The recommended length of this baffle is 3/4 inch, but this can vary a seventh either way without bothering anything. If the baffle is well-compacted but too short, the screen wire rectangle needs to be wider. If the baffle is too long, make the rectangle narrower.

To make the small baffle, the procedure is a little different. The rectangle is approximately 5 1/2 x 2 1/2 inches. The washer must be put on the bolt before spindling the screen around the bolt. If a piece of rod is used instead of a bolt, a couple of bumps of weld should be tacked onto one end of the rod to keep the washer in place. After spindling the screen around the hole, start the assembly into one end of the cylinder, making certain that the washer is inside the cylinder. If you form a baffle with the washer not all the way in place, the hole in the baffle will not be centered!

Another factor that determines how precisely the hole is centered is the length of the end fold that is recommended in order to eliminate loose ends inside the baffle (Figure 9). If this fold is not the right length, the hole will not be centered. The length that seems to work out best is whatever length constitutes a full circle of the hole size concerned.

Place the assembly vertically on the bench, with the bolthead down. Start the small-hole end of the piston down over the bolt and inside the piston. Strike the upper end of the piston with the rubber mallet. The strength of these mallet blows should be about the same as you would use to drive a large nail with a hammer.

Remove the small baffle from the die and check it for size and concentricity of the hole. Do not make this small baffle shorter than the recommended 3/8 inch. It can be up to 1/8 inch longer. If the baffles do not quite fill up the silencer tube, you can make up another thin baffle and press everything down into the tube just a little bit tighter, in order to make everything come out even. Do not leave more than 1/8 inch of empty space between the washer end cap and the top baffle. If you do, the force of the gases will push the baffles forward and you will have wasted space at the rear of the can.

When making the baffles for real, it is a good idea to put a small drop of solder on the end of the folds of the screen, both inside and out and on both ends of the baffle. To do this, just push the assembly to one end of the cylinder after forming the baffle and get the rod and washer out of your way. Use a soldering iron to melt a bit of solder onto both folds on that end of the baffle. Do not overdo this because every interstice that is closed up with solder is one less that can help contain the gases. Repeat this process on the other end of the baffle, by pushing the baffle to the other end of the cylinder. Once the baffle is so treated, it cannot become unraveled.

Now we start to get technical, taking up the manufacture of the can itself.

**WARNING:** Under federal law, any part unique to a silencer, such as a baffle, that has no other common purpose is an illegal part—*not* just the tube. Do not proceed further without the proper license.

## MAKING AND MOUNTING THE CAN

If you have access to a welder and a lathe, the tubing size of choice is 7/8 inch OD, 5 inches long for pistols and 6 inches long for rifles. The adapter is turned out of solid rod and is welded into the end of the tube (Figures 32, 33, 19, and 18).

If you do not have access to the above tools, you can still make the can out of PVC tubing and the aforementioned Acraglas epoxy to mount the 3/4 x 24 bolt in the end of the PVC tube (Figures 26 and 27). To further solidify this mount, insert three 1/16-inch roll pins into holes drilled through the sides of the tubing and into the sides of the bolthead. The maker must decide which method to use and then follow the corresponding directions and dimensions for that type of tube.

Whichever type of tubing is used, the rear end of it must be cut exactly at 90° to the line of the tube's long axis. This is easier to do with a lathe but can be done without one. A miter box can be used to get a square cut on the PVC tubing, and so can a tubing cutter. Use a square to check this out because the adapter (or bolt) will be mounted flush with the rear edges of the tube. If the tube is not true, the adapter will be installed crookedly, and the can will not be in line with the bore.

If welding the adapter into the tube, cut a bevel on the inside of the tube and on the edge of the adapter to allow good penetration of the weld. The face or shoulder of the adapter must be clean, i.e., free of excess weld or other obstruction to the shoulder coming up squarely against the muzzle of the gun (Figure 18).

V-type threads—as opposed to Acme, square, or buttress threads—only hold things on; they do not necessarily hold them straight. In our application here, where the bore of the can must align with the bore of the gun, the components are held in line by setting up a shoulder on the male part for the matching shoulder of the female part to come up against. In order for there to be enough play in the assembly to allow the shoulders to square up against each other, a so-called "empty thread" must be cut into the male part. This empty thread is a groove that is cut just in front of the male shoulder. This groove must be a little bit deeper than the threads and a little bit wider than the distance that two threads cover (Figure 33).

When making the adapter out of solid stock, first turn the OD to a close fit in the end of the tube. Then make a plunge cut to form the threaded portion of the adapter and cut the adapter off the stock. Do not cut the part to be threaded to its finished diameter before welding it into the tube. Use ordinary hot- or cold-rolled steel to make this adapter; do not use any grade of tool steel because it can harden under the heat of the welding process.

Leaving the smaller part of the adapter a bit oversized will allow you to straighten up any misalignment that occurs in the welding. To do this, just chuck the tube in the lathe, with the piston inside it for support, and take a light cut along the area to be threaded.

While the tube is still in the lathe, center drill and finish drill the hole through the adapter and thread it. For convenience in threading it, just use a die and die holder supported by a live center in the tailstock and turn the headstock by hand. As the die advances, just turn the tailstock to match, keeping the die holder supported by the live center.

Keeping the tube in the chuck while all these operations are done will ensure that everything is straight, relative to the axis of the tube. This is also a good time to cut off the excess weld on the adapter and the tube and to cut the empty thread.

If using the PVC tubing, use a 3/4-inch-long, 3/8-inch-fine thread bolt as the adapter. The thread on this bolt is 24 threads per inch (TPI). This bolt is driven head first into the rear end of the tube. Use the vernier to scribe a line around the end of the tube one-half the thickness of the bolthead from the end of the tube.

Looking at the bolt from the end, determine where you want to drill the three 1/16-inch holes for the pins that will help to hold the bolt in the tube. Center punch and drill these holes through the PVC, going far enough to mark the side of the bolthead. Repeat this operation on the other two locations (Figures 26 and 27). In the next step, you must put the bolt back in the same place it was when you marked it because the holes will be different enough to cause problems. Therefore, it is a good idea to mark the tube and the bolt now with a Magic Marker to ensure that you can get everything back the way it was.

Remove the bolt from the PVC and clamp it in a vise. Center punch and drill the bolthead for the three 1/6-inch holes. Mix up some Acraglas, smear the sides of the bolthead with the epoxy, and return the bolt to the end of the tube. Then drive the three 3/8-inch-long roll pins into the holes. The Acraglas requires 24 hours to harden fully, so do not allow the assembly to be disturbed in any way during that period, or you risk disturbing the alignment. The 1/4-inch hole and the empty thread must be done first. If not using a lathe, use the method explained for a brass pilot piece.

Now you are at the critical stage. You must counterbore and thread the barrel (Figure 20). If you don't get this right, the result is a ruined barrel (unless the barrel is long enough to permit the bored-out portion to be cut off and redone).

If you have never tapped a hole before, it will be necessary for you to gain some experience at this process before risking a (perhaps irreplaceable) barrel. First drill a clean hole with the piloted drill bit. You then need the 3/8 x 24 tap and a so-called "T-handle" for that tap. A good tapping lubricant is nice to have, too, but I've gotten by with using WD-40.

There are three kinds of tap for any given hole size: the taper, plug, and bottoming taps. The taper tap is used first because its longer lead lets us start the thread-cutting action more easily. The plug tap then follows, and the bottoming plug is last. This is the ideal setup, but most holes are just threaded with the plug tap. On a blind hole (one that doesn't go all the way through the material), the bottoming tap is necessary if the hole must be threaded as close as possible to its bottom. If you are limited by a short pistol barrel and thus want to use a shallow counterbore, the bottoming tap may be called for.

To start the tap into the hole, cup your palm over the center of the T-handle and take about a half-turn (clockwise) into the hole. Repeat this procedure a couple of times or until the tap will stand up in the hole of its own accord.

Now is the time for you to squirt in a bit of lubricant and to check the tap for straightness in the hole. This has to be done mainly by eye, although a square may be of some help. Remember to check it in two planes, 90° apart, or you may get it right one way and be off in the other.

Turn the tap further in but be sure to back off a partial turn for each partial turn that you advance. This backing off allows the cuttings to fall free from the cutting surfaces of the tap. If you just force the tap on into the hole, it will eventually clog with chips and either break or cut a crooked thread. I hope I have emphasized that the straightness of a can's mount is of paramount importance.

After tapping the hole all the way to its bottom, remove the tap and clean the chips out of the hole. A compressed-air nozzle is ideal for this job. If one is not available, just invert the barrel and tap on its sides with a mallet. The chips may also be flushed out with solvents. If the tap has thrown up any burrs on the crown of the muzzle, they must be removed or they will not let the shoulder of the adapter come up against the muzzle squarely. This meeting of the two surfaces is what determines the alignment of the can with the bore, so get it right!

If, after screwing the can onto the barrel, it is obvious that it is not straight, you will have to make a choice: either discard the can and build another, or attempt to straighten it. Before doing either, though, be sure that the fault lies with the can. Screw a bolt into the barrel and check it for squareness. If it is not straight, you must either shorten the barrel and cut new threads or replace it.

If you suspect that the mount is crooked, wait for an objective test to tell you if it truly is before making any rash decisions. By this objective test, I mean firing the piece with the can in place and observing the results. I will discuss this test later, after having run through the procedure for packing the can.

## FINAL ASSEMBLY

You've got the can mounted on the barrel and you know how to make the baffles, so let's proceed with the final assembly. As I said before, some guns require that the baffle directly in front of the muzzle have a large hole, and all cans are more efficient if this is the case, so we will first install a large baffle. In fact, we will install two, except in the case of those using PVC tubing, which has no large baffles at all.

Make a large baffle and solder the folds on both ends to eliminate raveling. Start it into the main silencer tube and push it to the bottom with a piece of wooden dowel that is larger in OD than the hole in the baffle, yet is smaller than the ID of the can. Repeat this, so that you have two large baffles at the bottom of the can. Next, make a small baffle and push it up against the two large baffles. Then install one of the inner-tube rubber disks.

### *Rubber Disk Wipes*

I must digress here to discuss making these disks. Cut open an old inner tube and lay it out on your bench with the gray inside surface facing up. Use the piston from the die set as a template to make the disks the right size. To do this, just place the piston vertically down on the rubber and use an ink pen to draw a circle around it. Repeat this three to 11 times, depending on whether you are using PVC or not.

For the PVC version, the baffles are all 3/4-inch long, and each one will have a rubber disk (or wipe) in front of it. The 7/8-inch steel tubing version will have a wipe in front of each small baffle only. If you put a wipe in front of the large baffles, you will lose the effect of exposing more baffle surface, which was the whole idea behind using two sizes of baffle in the first place. Remember, do not attempt to make the hole in the wipe at this point. That job is better taken care of after the can is fully assembled.

After seating the first wipe in the can, continue putting baffles into the can, alternating between the small and large types. In other words, first a large baffle, then a small baffle, then another wipe, and so on until the can is completely packed (Figure 23).

### *Snap-Ring Groove*

I have left the cutting of the snap-ring groove till last for a reason. Although it can be done while the tube is in the lathe, using the special lathe bit shown in Figure 22, the baffles have a tendency to get caught on the edge of the groove as one tries to get them into the can. If using a lathe, one can wait until the next-to-last baffle is in the can before cutting the groove.

If using the PVC tubing, the groove can be cut with the mandrel-mounted circular saw accessory sold for use with the Dremel Moto-Tool hand-held electrical grinder. Because the PVC is so soft, the much slower RPM of your electric drill will suffice to make this cut, so a Dremel is not necessary. However, once you discover these wonderful little worksavers, you will never want to be without one again.

The groove should be cut with its front edge about 1/8-inch back from the front of the can (Figure 17). It should be just a little wider than the snap ring is thick, or about 0.100 inch. It should be between 0.020- and 0.030-inch deep and should have a sharp corner at its front so that it offers the maximum amount of resistance to being blown out of the can by the pressure of the powder gases.

It does not matter whether the last baffle is a small one or not. Just be sure to cover the last baffle with a wipe before inserting the end-cap washer and the snap ring (Figure 17). Ta-da! Finished silencer! Now, to test the product.

## TESTING THE SILENCER

Testing the can must be done with full attention paid to all details. A bad decision here will cause you to throw away a can that could have been saved or, worse, to use a can that isn't right.

First, be sure that the threads are snug. Do not just step outside the door of your shop and fire a shot into the air. If something is amiss, the washer and snap ring will go flying, and you will not have the slightest idea what is wrong. Instead, put a piece of 2x4 board edge-up on the floor and fire down into that. Have only one round in the gun at this point so that you will not be screwing or unscrewing the can at the risk of your hand. It's easy to get caught up in the task at hand and forget basic firearms safety, so watch out!

Place the end of the can down on the edge of the 2x4, with the gun held just as vertical as possible and centered on the board. Fire the shot and carefully raise the gun a few inches. Move the gun down out of your line of sight, while holding everything relative to how it was when you fired. Look at the hole. Is it nice and round, or is it egg-shaped? Is the area around it clean, or are there splatters and stains of lead fragments and powder residue? Of course, if the washer and the snap ring are lying there on the wood, you know that something is way off.

Sometimes you can tell by the hole (or by smears on the washer) which way the can is out of line. Remember my saying to hold everything relative? If you just cavalierly move the gun away after the shot, how will you know what's what? Let's say that you have determined that the can is out of line to the left. It is possible to correct a misaligned can if it is not too far off.

## Correcting Misalignments

I've brought cans back into alignment that were off by a quarter of an inch, and I've wallowed-out the baffle holes in order to save cans that were farther off than that.

The steel tube can be more easily straightened than the PVC version because steel tubing will bend and because the epoxy bond can't stand much flexing, the way a weld can. Hold the gun-and-can assembly vertically and administer a light rap with the rubber mallet on the same side as the way the can is misaligned. The tubing will bend and the threads will stretch rather easily, so don't bash the thing too hard. You can always tap it again if it turns out that the blow was insufficient to the task. But you cannot get away with doing this again and again indefinitely, so it is very important to move the can in the right direction and not to go too far. After each blow is administered, it will be necessary to retighten the can on the gun because the threads will have stretched.

If the can was misaligned badly enough, the slug will have damaged some of the baffles. Any damaged baffles will have to be replaced before any more testing is done, because they might cause you to make the wrong decision about the can's condition. I test for this by running a 9-inch piece of drill rod through the can. This rod should be the same OD as the small rod in the baffle-forming set. If the rod passes through the can with minimum trouble, you can be reasonably sure that the bullet's pathway is not blocked, allowing for the fact that the wipes will present some resistance. The front end of this rod should have a rounded surface, so that it is not prone to tearing the wipes as it passes through them.

I should stress that you must not strike the side of the PVC tube. The blows must be delivered to the end of the tube, where the adapter is, directly in front of the gun's muzzle. This may require that the gun be clamped in a vise and a piece of hardwood be used as a drift punch to transfer the force of the hammer blow directly to the threads without stressing the epoxy joint between the bolthead and the PVC tube.

### *Burning Holes in the Wipes*

Once a successful firing indicates that all is well, it

is time to burn out the holes in the wipes, so that they will impact on the bullet in a uniform manner. Heat the rounded end of the 9-inch long, 1/4-inch drill rod red-hot with a torch or on the kitchen stove. Pass this heated end through all of the wipes, pausing a bit at each one if you want to burn the hole larger.

There should be no reason that silenced fire is not fully as accurate as what the gun and ammo combination is capable of when the silencer is not mounted. Check this out. If the size of a 5-shot group on the paper is significantly and consistently larger with the can than without it, something in the can is affecting the flight of the bullet. Do not base a decision on two or three shots, however, because they could be the result of shooter error. Also, remember that the attachment of the can will usually change the gun's point of impact, so do not feel that something is wrong with the can just because the gun shoots higher or lower, or whatever, when the can is in place. If the group size is the same, all is well. You just must rezero the sights to reflect the new point of impact.

Remember what I said earlier about dedicating guns to silenced use? You can save about $15 in the construction of this can by modifying a set of kid's paper scissors to serve as snap-ring pliers. Start with the pointed kind and use either a Dremel tool with carborundum disk or a three-cornered file to make the necessary little nipples on the scissors that allow them to fit in the two holes of the snap ring (Figure 35).

If it becomes necessary to remove the baffles from a tube, firing a round without the snap ring in place will usually displace at least some of them. To get the others out, you need a long hook of some sort. This hook can be made on the other end of the wipe-burning rod. Just grind a long, tapered point on the rod first and then grind two opposing flats on this tapered point. Use pliers or a vise to bend the point around into a hook shape, with the flats on the curved surfaces. This tool can now be used to reach into the central hole, through the central hole in each baffle, and draw it out of the tube. This process ordinarily destroys the baffle, so do not expect it to emerge unscathed!

It is possible to conceal a can inside any .22 rifle barrel's configuration. Remington once marketed a "gallery rifle" in .22 short based on Browning's little autoloader with silencing chambers machined internally in the muzzle end of the barrel. The barrel is simply shortened, and a can that matches the barrel's size and shape is threaded on. The joint can be hidden under a barrel band (a piece of metal banding that is wrapped around the forestock and the barrel). It is also possible to silence the various full-auto conversions of the 10-.22 and other popular .22 rifles on the market. Just contact the source of this manual for more advanced models.

## SHOULD YOU VENT THE BARREL?

You may have been under the impression that it is necessary or desirable to vent a barrel (i.e., bore holes into it radially in order to have an efficient can). This is not true. A vented barrel can never match the gas-release rate of a barrel that is cut off just at the point where the vent holes would have begun. Drilling holes into a barrel is quite a bit of extra work, and the drill throws up burrs inside the bore that must be removed to prevent inaccuracy. This requires either reaching in through the hole with a special little hook-shaped deburring knife or reaming out the barrel from the muzzle back to the vents closest to the breach. The bullet is deformed a bit when it has to slide past these vents, so they can be expected to hurt accuracy somewhat. If the reaming method is used, the extra barrel becomes useless because it has no rifling and lets gases slip past the bullet.

There are a couple of exceptions to the above objections to venting barrels. If one wished to retain the legal length of 18 inches of barrel on a .30-06 hunting rifle, for example, and yet have the minimum OAL when a can is mounted, he might vent the front 4 inches of the barrel and then ream out that much of the rifling. Then, about 8 inches of the can could be telescoped back around the barrel and a double-mount contrived so that the can would have two working chambers. In other words, instead of having 14 inches of can in front of the 18-inch barrel, he could have only 10 inches, because venting the gases into the rear chamber would have relieved some of the heat and pressure that would otherwise have to be handled by the front (normally packed) chamber.

The other exception occurs on some submachine guns (notably the Heckler & Koch), wherein it is desirable to vent the barrel in order to reduce the velocity of the 9mm ammo to the point that it is no longer supersonic. Since most of the H&K's rather long barrel is not out where we could get at it to simply shorten it, venting is the only recourse left. Of course, a man who reloads could simply use the 160-grain slug designed for the .38 Special (sized .355 instead of the usual .357) and drive this heavier slug to 1,000 fps instead of the 9mm's normal 125-grain bullet at 1,300 fps. The issue of vent holes distorting the bullets is not very critical on a burp gun, especially with full-jacketed ammo.

The tiniest hole in a weld or in the epoxy will allow gases to escape. If the can is noisy, use soapy water to check for leaks. Smear the joint with a thick coating of the soap solution and fire one round while looking for bubbles.

Even the slightest movement of a sight will cause a

significant change in the bullet's point of impact, and threads will stretch a little bit every time they are snugged up. For this reason, mounting the front sight on the can is not a good idea, unless the can is to be at least semipermanently mounted with Loc-Tite thread-seating compound. A scope is a big help when working with rifles, as it gives a line of sight higher than the bore.

## SUMMARY

There is quite a lot of nonsense in print about making silencers. Some of what is available is merely the patent drawings of the old Maxim design, with its need for special punch press dies. Others require the cutting of very coarse threads into a large piece of aluminum rod. Such a rod is expensive, and cutting such deep threads requires multiple passes to be taken to form them on a lathe. Spiral baffles are inefficient because such cans are huge. Other cans use nothing but faucet washers.

You may have been under the impression that some sort of vented tubing is required for silencer construction. This idea arose out of the fact that steel wool burns out so quickly, and the vented tube is a way of getting the steel wool to last for more than two or three shots. This works, more or less, but it requires that the can be significantly oversized, compared to what it could be if made from the proper materials.

Good luck, good shooting, and be quiet!

# Part Three

# Illustrations

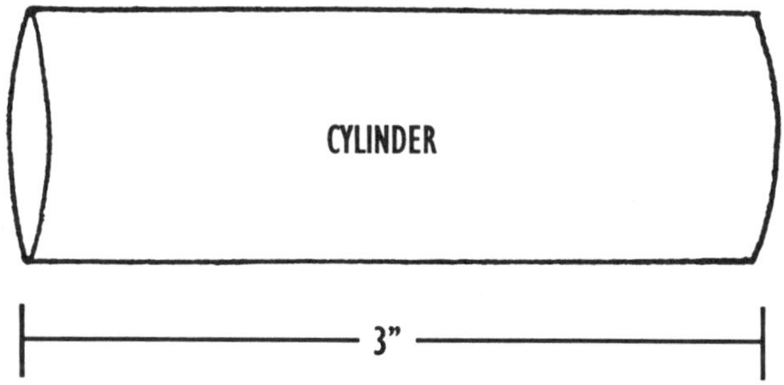


**Figure 1**

Cylinder can be either steel or PVC tubing. If PVC is used, it should be 1/2" inner diameter. Use .065 wall steel tubing 7/8" outer diameter. Deburr inside and out.

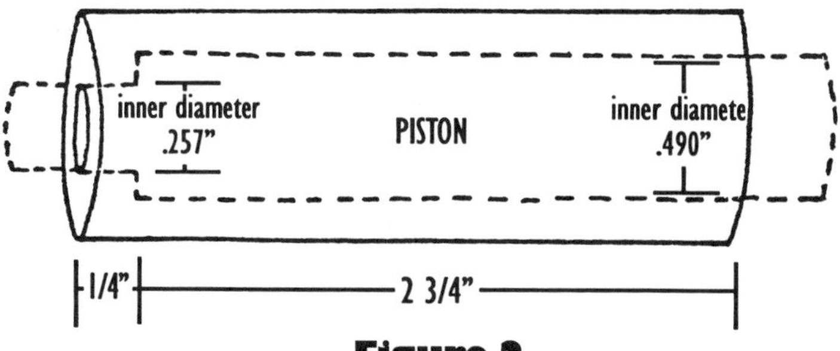


**Figure 2**

If PVC is used, outer diameter of piston is .490" If steel is used, the outer diameter should be .740" to match the .745" inner diameter of the .065 wall of the cylinder. The counterbore is unnecessary for the PVC version.

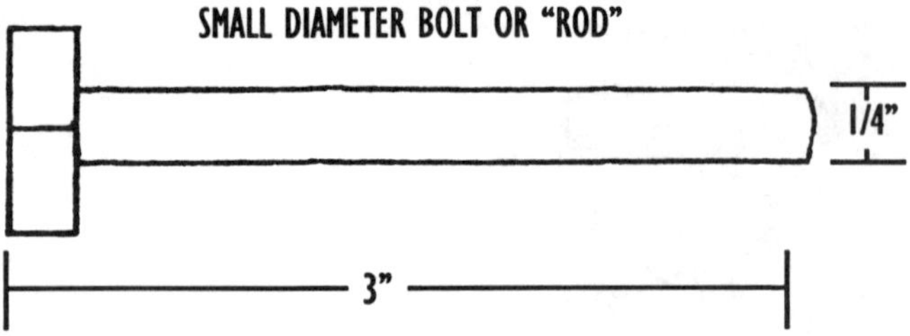


**Figure 3**

This large diameter rod is optional. It is not used on the PVC version.

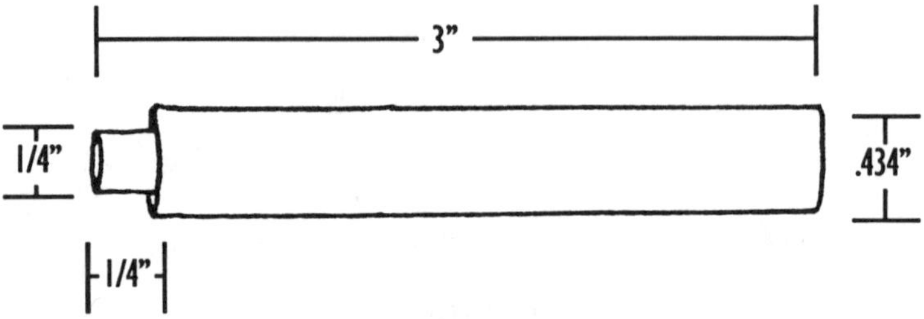


**Figure 4**

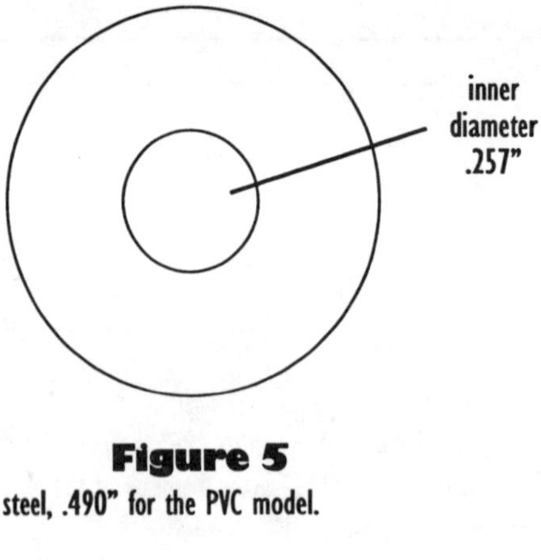


**Figure 5**

Outer diameter .740" in steel, .490" for the PVC model.

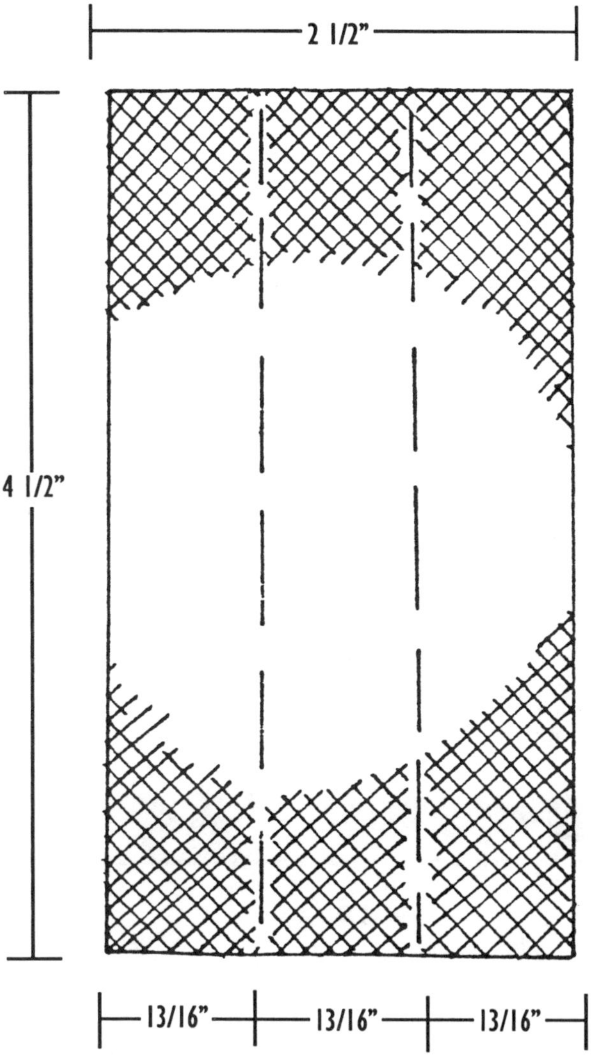


**Figure 6**

Screenwire rectangle, showing where to fold. Sizes are approximations only. Use bronze screen and anneal it before attempting to form baffles.

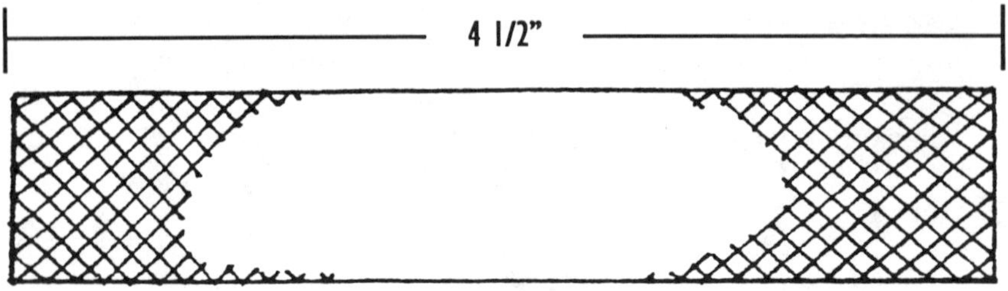


**Figure 7**

Screenwire, folded into thirds lengthwise. Rough edges not shown. Spindle this "clean" side toward the rod.

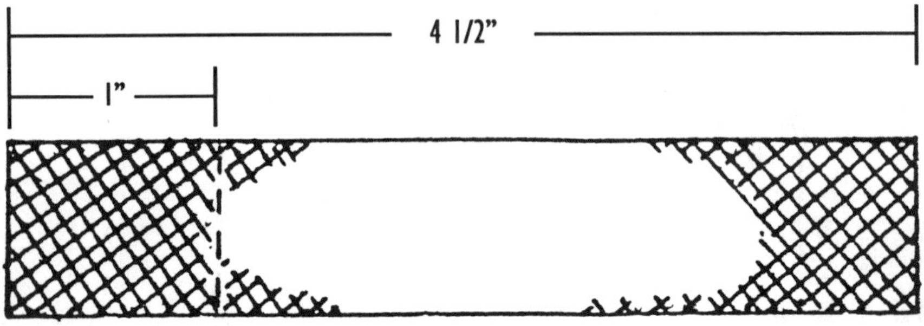


**Figure 8**

Screen rectangle, showing end-fold.

**Figure 9**

Fold the rough edges toward each other.

**Figure 10**

Spindle the screen around the rod starting from the fold with rough edges toward the outside.

**Figure 11**
End view of small baffle.

**Figure 12**
End view of large baffle.

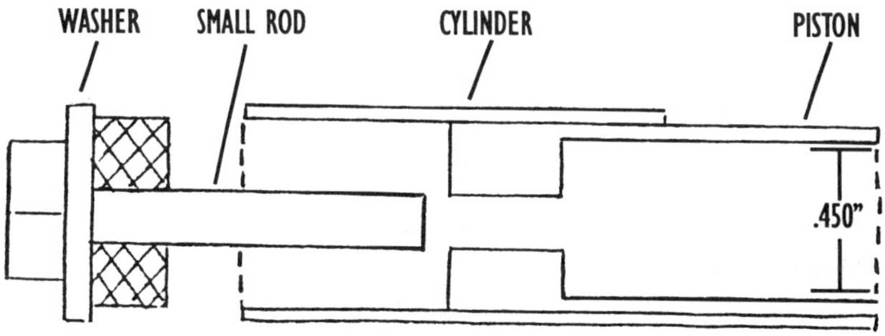


**Figure 13**
Die assembly after making small baffle.

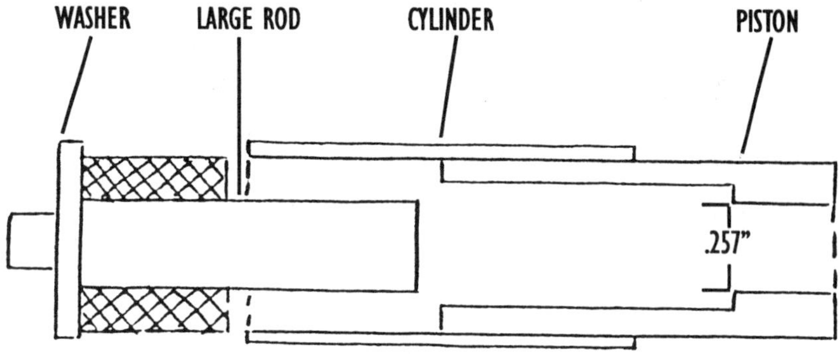


**Figure 14**
Die assembly after making large baffle.

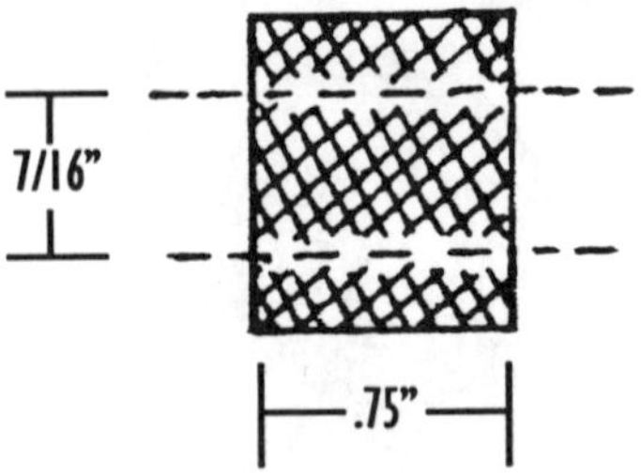


**Figure 15**

Side view of large baffle.

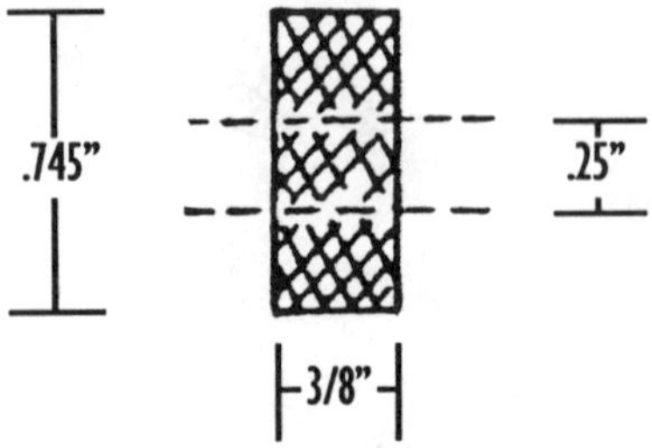


**Figure 16**

Side view of small baffle.

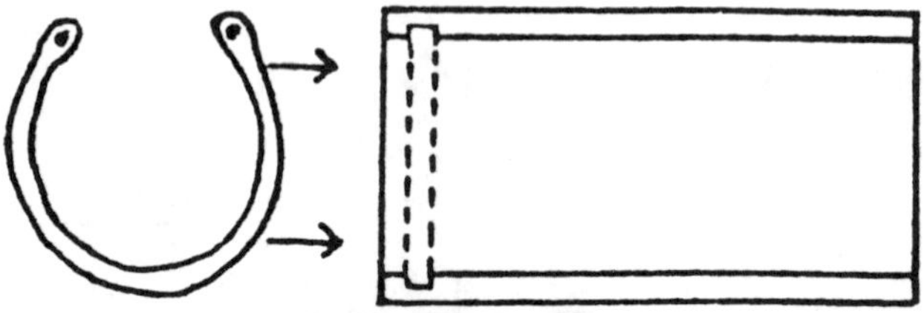

**Figure 17**

Main tube showing snap ring groove. Groove must be cut .025" back from the front edge of the tube. Make the front edge of the groove square so that it will hold the snap ring securely. Groove must be .020" to .030" deep and wide enough to accept the snap ring.

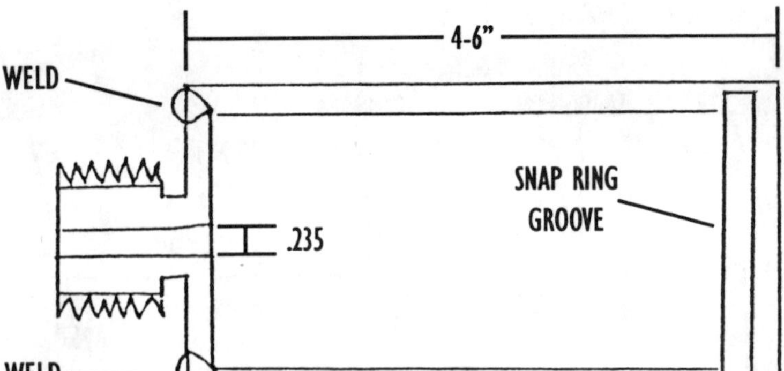


**Figure 18**

Main tube showing adapter welded into rear of tube. Deburr inside and outside. Cut off excess weld so that rear "shoulder" is a "clean" surface to mate with the muzzle.

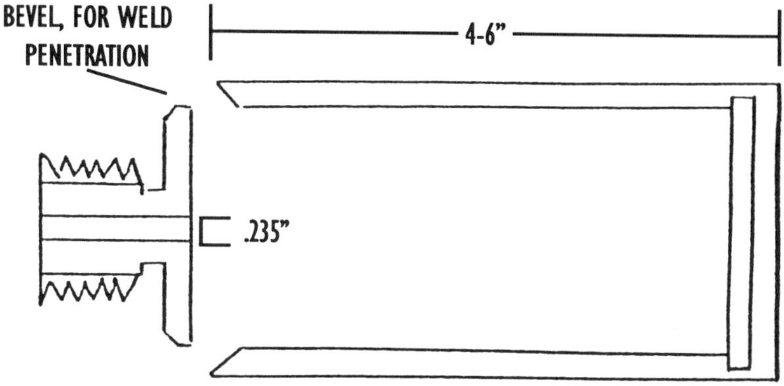


**Figure 19**

Rear end of tube must be cut precisely "true" at 90° to the long axis of the tube. The adapter must be mounted straight.

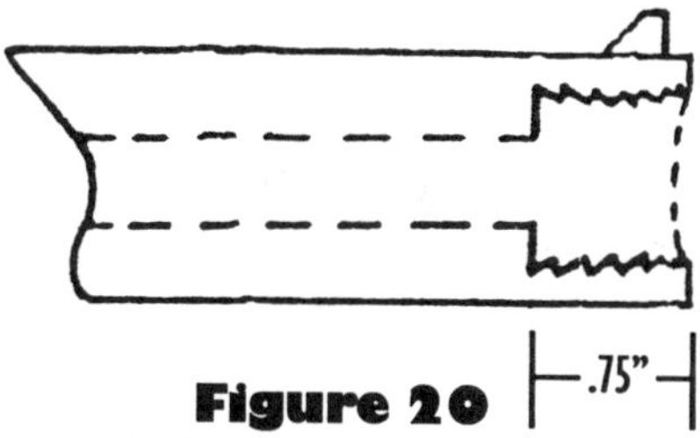


**Figure 20**

Muzzle of barrel showing 3/8" x 24 thread.

**Figure 21**

When checking the angle of the cutting lips of a drill bit, observe the face of the lips.

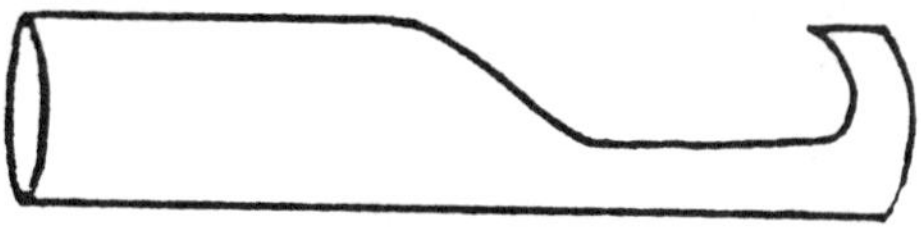

**Figure 22**

A 1/2" lathe bit shown ground as a "cutter" for making the snap ring groove in the main tube. Note: the end of the hook should be sharp so that the groove will have a square corner.

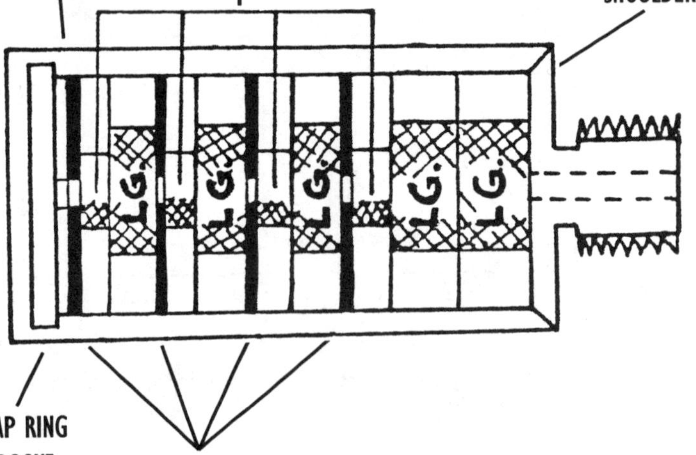


**Figure 23**

Finished "can" shown in cutaway view revealing the sequence of large baffle, small baffle, and wipe.

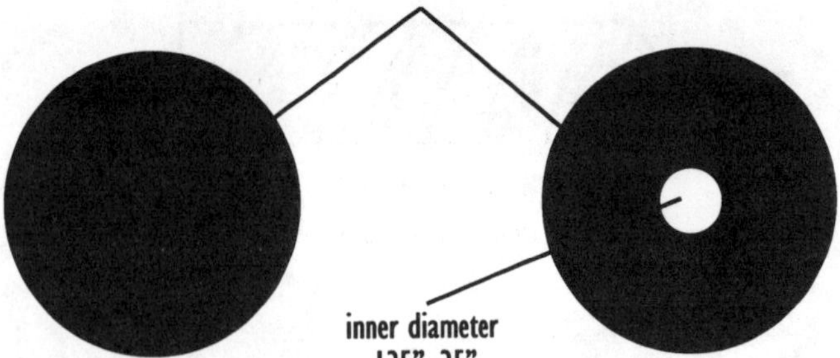


**Figure 24** **Figure 25**

Rubber disks or "wipes." Do not punch holes before installing wipes in the can. The outer diameter of these pieces of inner tube should be .5" for the PVC model.

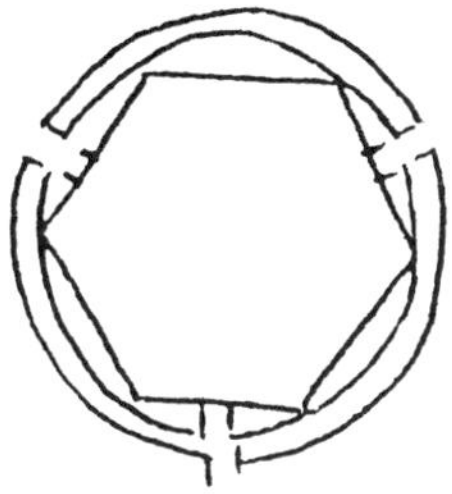

**Figure 26**

If using PVC pipe for main tube, for lack of a welder, drill three 1/16" holes through the sides of the tube and into the sides of the bolthead after putting in the "trued" end of the tube.

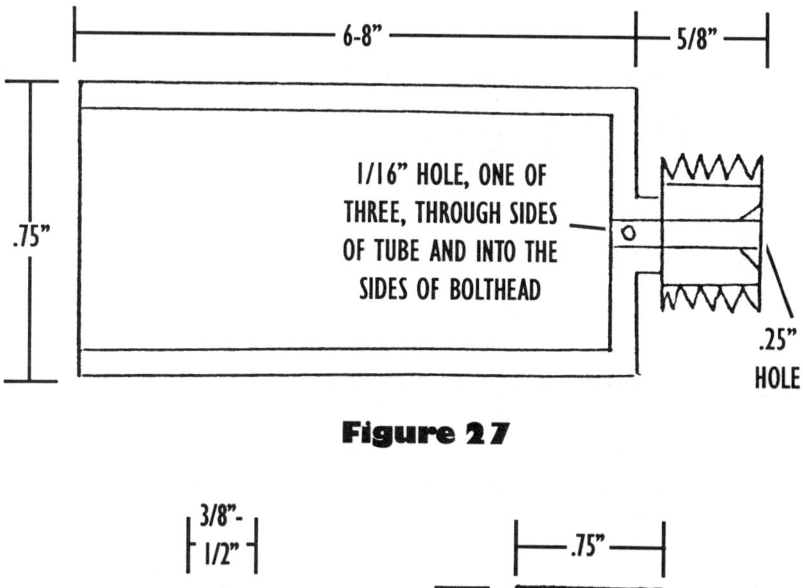


**Figure 27**

**Figure 28**

End view of brass pilot.

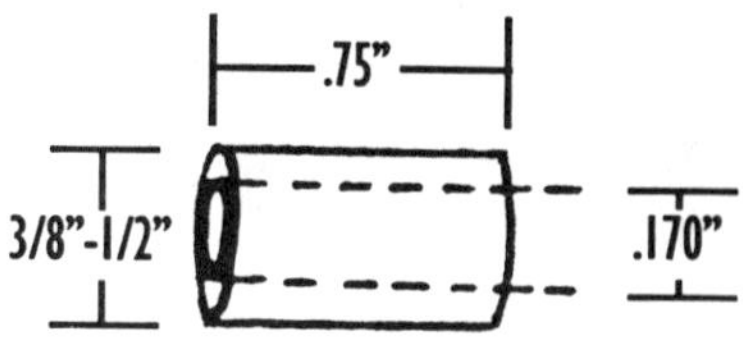


**Figure 29**

Side view of brass pilot.

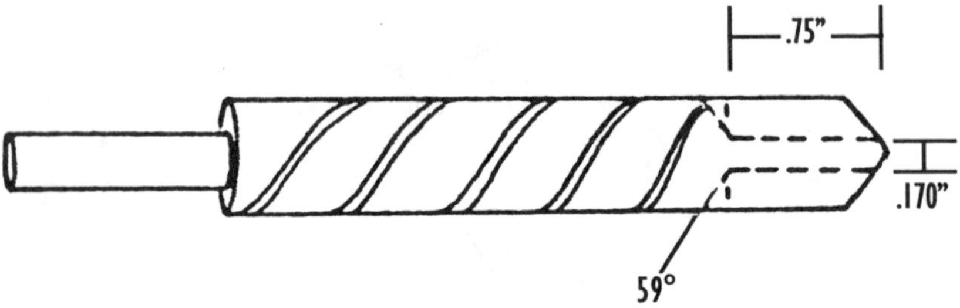


**Figure 30**

The tap drill for 3/8" thread is a letter "Q" bit. This bit is shown with the "nipple" ground on its front end. Remember that the chip clearance must reground on the 59° cutting lips.

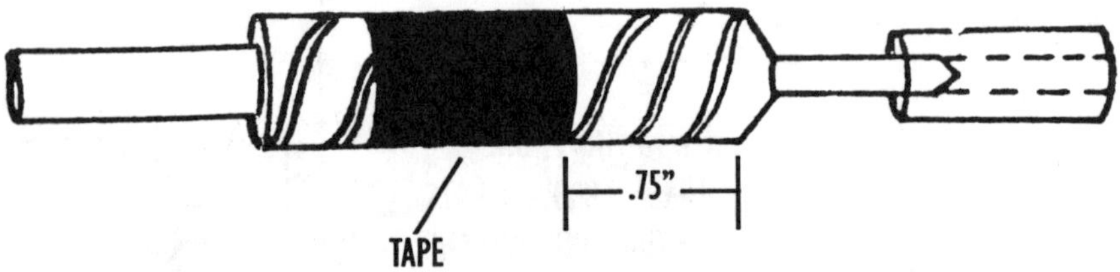


**Figure 31**

Reduce outer diameter of pilot to slip-fit in barrel after silver-soldering it on. Bit is wrapped with tape to mark depth of counterbore in barrel.

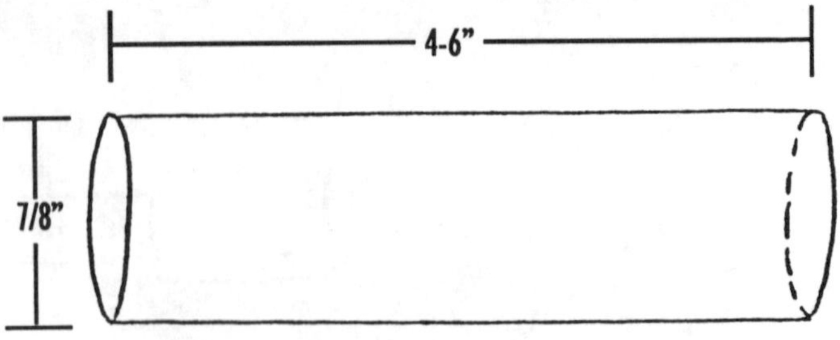


**Figure 32**

Main tube. If using steel, it is 4-5" of .065 wall tubing to match the cylinder of the baffle-forming die. In PVC, it is 6-8" long.

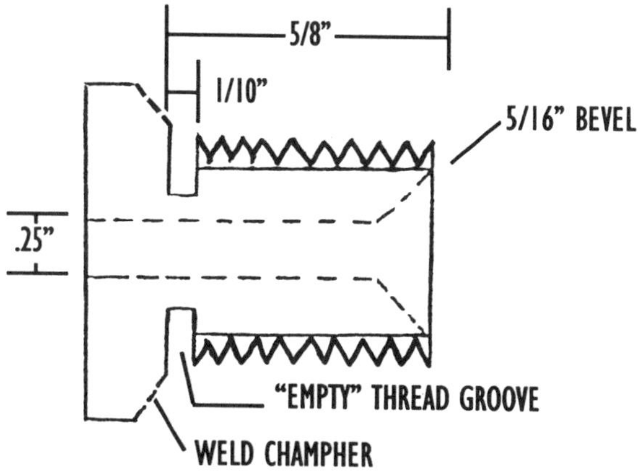


**Figure 33**

Adapter to be welded into rear of main tube. "Empty" thread groove is cut deeper than the thread.

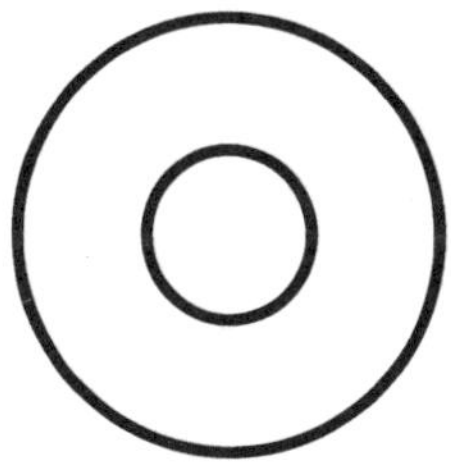

**Figure 34**

Endcap washer. Outer diameter is .730" for steel and .490" for PVC. Inner diameter is 5/16".

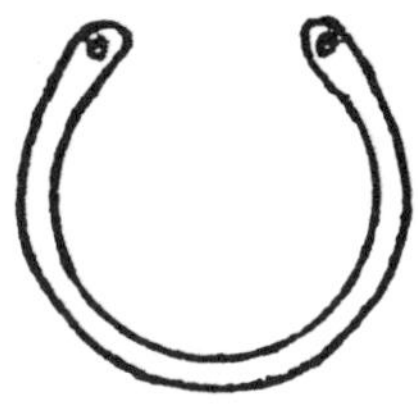

**Figure 35**

Internal snap ring. 13/16" in steel version and 9/16" for PVC. The ends of a pair of pointed paper scissors can be ground to fit the holes in this snap ring.